Controlling exposure to coating powders

HSE BOOKS

First published 2000

ISBN 0 7176 1761 0

This guidance is issued by the Health and Safety Executive. Following the guidance is not compulsory and you are free to take other action. But if you do follow the guidance you will normally be doing enough to comply with the law. Health and safety inspectors seek to secure compliance with the law and may refer to this guidance as illustrating good practice.

Contents

Introduction

1 This guidance document provides advice on the assessment and control of exposure to coating powders. Case studies of companies who have made improvements to their plant and / or working practices are also included to illustrate what can be achieved. This guidance also contains a training module on CD (see paragraph 97).

Note: Some of the photographs in this booklet show operators wearing a respirator and/or gloves whilst others do not. This is not to suggest that suitable and adequate personal protective equipment has been chosen in each case, or that it is being worn correctly. See paragraph 83 for advice about the use of this equipment.

Who is this booklet for?

2 If you run or manage a business which uses coating powders then this booklet should help you. If you are a safety representative or if you have other health and safety responsibilities within the business, you may also find this booklet helpful.

What does this booklet do?

3 This booklet helps you to manage the risk to the health of your employees from exposure to coating powders in the same way that you manage other parts of your business. It shows you the steps that need to be considered in assessing and controlling exposure, but also demonstrates through case studies and simple principles that the answer is often straightforward once you have taken a fresh look at the process. It demonstrates that failings that result in increased worker exposure may also be reducing the profitability of your business.

4 It provides employers with details of the latest available coating powder plant technology. Information on technological developments is provided without any comment on the relative merits and without endorsement of these systems. This information should be viewed in the context of all the guidance in this document and what can be achieved through following good practice. For further information on these systems you should contact your equipment supplier or trade association (see Appendix 1).

Implementing control measures to reduce worker exposure can reduce running costs.

What if I still need help?

5 The aim of this guidance is to raise understanding of how workers are exposed to dust from coating powders and how this can be controlled. It describes, in general terms, the operating factors that affect operators' exposure and that can also lead to increased plant running costs. **More detailed technical information on coating powder technology can be obtained from some of the publications and organisations listed in Appendix 1.**

How to use this guidance

6 This guidance provides advice on controlling exposure through the steps needed to assess your workers' exposure and through comparison with best practice. It is important that you understand how exposure arises in your workplace; however, it is equally important that you consider whether you follow best practice. If you decide after reading this guidance that occupational exposure is being adequately controlled, it may still be possible to do things better to further reduce this exposure. In some cases, for example where exposure is to a carcinogen or a substance assigned a maximum exposure limit (MEL), it is a legal requirement to reduce exposure as low as is reasonably practicable (see paragraphs 14 and 16). Such improvements may also make your business more efficient through reduced costs and improved quality as well as reducing occupational exposure.

How does electrostatic powder coating work?

7 Industrial coating powders are produced by first blending together pigments, resins, curing agents and other additives. The premix is extruded, during which it becomes molten, and has to be rapidly cooled before being milled to form discrete particles. The resultant coating powder is applied to the work-piece, via electrostatic or tribo application systems.

8 The charged powder particles carried in the airflow from the gun are attracted to earth via the work-piece and deposited onto it. The thickness of the powder deposited is limited by the fact that the deposited powder then acts as an insulator and progressively decreases the attraction of the additional powder. Whether a particular particle of powder 'sticks' to the work-piece is a balance between the level of charge on the particle, the earthing of the work-piece, the insulation effect of deposited powder and the force of air used to carry the powder from the gun towards the work-piece. A point is reached where additional spraying activity will not result in an increase in the amount of powder deposited, as the attraction of the

powder is less than the force of the air which tends to blow the powder off the work-piece as fast as it is deposited.

9 Application can be via automated or manual systems, with the work-pieces transported by an overhead conveyor through a spray zone containing the automated or manually operated spray guns. Once coated the work-piece is then transported through a stoving oven. In smaller trade coaters the process may be fully manual with individual items placed inside a spray booth for coating.

10 The majority of coatings powders used in the UK are thermosetting and are either:

- epoxy;
- epoxy/polyester;
- polyester based; or
- other powder types including polyurethane and modified nylons.

11 Epoxy coatings generally have good chemical resistance, but tend to chalk in sunlight and, therefore, are usually used for indoor applications. Polyester coatings are applied where exterior use is the prime consideration as they are very durable and resist yellowing and chalking when exposed to UV. Epoxy/polyesters are hybrid materials which display good heat resistance and general performance properties. They are suitable for interior applications as the epoxy component will still chalk on exterior exposure.

The health effects of coating powders

12 Workers may come into contact with coating powders via the skin, and also by inhalation. A fraction of the inhaled material will reach the back of the throat and be swallowed, and there may also be some oral ingestion if employees do not use a high standard of personal hygiene. The curing/hardening agents used in coating powders are reactive materials, and may therefore cause direct irritation of the skin, eyes and respiratory tract or allergic skin reactions. In addition to such local site-of-contact effects, some inhaled material may be absorbed into the bloodstream and may cause toxicity at other sites within the body. There is concern that some curing agents may cause asthma. Another particular concern about some curing agents is the potential to damage genetic material, a cause of some diseases including cancer and impaired fertility.

13 One curing agent of particular concern, which is used in polyester powders, is TGIC (triglycidyl isocyanurate). TGIC has been classified under the Chemicals Hazard Information & Packaging for Supply (CHIP) Regulations (1994) (as amended)[1] as Toxic, Irritant and as a Category 2 Mutagen. It has been assigned the following risk phrases for labelling purposes:

R23/25	Toxic by inhalation and if swallowed.
R41	Risk of serious damage to eyes.
R43	May cause sensitisation by skin contact.
R46	May cause heritable genetic damage.
R48/22	Harmful: danger of serious damage to health by prolonged exposure if swallowed.
R52/53	Harmful to aquatic organisms, may cause long-term adverse effects in aquatic environment.

Legal requirements

14 The Control of Substances Hazardous to Health Regulations (COSHH) 1999[2] impose duties on employers to assess the risks to health arising from exposure to hazardous substances, and to ensure that exposure to these substances is prevented or, where this is not reasonably practicable, adequately controlled. The guidance and case studies included in this booklet give examples of control measures that you may find useful when considering how to comply with this duty.

15 The Environmental Protection Act 1990 (EPA) requires employers to control the risks to the environment from pollution arising from certain industrial and other processes. Industry has to comply with limits for emissions to the environment. Many of the controls implemented to comply with this legislation may have an added effect in reducing exposure to coating powders in the workplace.

Occupational exposure limits (OELs)

16 OELs are limits set by the Health and Safety Commission and have legal status under the COSHH Regulations. A list of current OELs may be found in *EH40 - Occupational Exposure Limits*[3] (revised yearly). There are two types of limits, Occupational Exposure Standards (OESs) and Maximum Exposure Limits (MELs). These limits represent a concentration of the hazardous substances in the air, averaged over a reference period of time (8 hours and / or 15 minutes). An OES is

set at a level that will not damage the health of the workers exposed to it by inhalation day after day (based on current scientific knowledge). MELs are set for substances which may cause the most serious health effects such as cancer and occupational asthma, and for which safe levels cannot be set or where safe levels may exist but control to those levels is not reasonably practicable. In essence, for OESs, exposure must be reduced to the level set and for MELs, exposure should be reduced so far as is reasonably practicable and in any case below the MEL.

Case study 1

Peateys Coatings are a trade coater who use both wet paint and powder coating technologies. They were concerned about their workers' exposure to dust and their level of awareness of the work and associated risks. To improve the working environment Peateys have carried out the following:

- ***Improved ventilation.*** The cyclone filters were removed and the main filters used to remove the dust. This increased the airflow into the booth openings. Cyclones filters are generally used for powder reclaim. Peateys do not reclaim over-sprayed powder and so the cyclones were not needed (see note below).
- ***Improved culture.*** Co-operation within the company has been improved by encouraging workers to report defects and plant problems which are acted upon quickly.
- ***Training.*** All spray operators received training from the equipment suppliers. New sprayers work alongside experienced sprayers until proficient at the job. Operators are now graded for their spraying proficiency.
- ***Process changes.*** Operators were found to be leaning into the booths and lifting their visors to see, when spraying black powders. This was found to be due to insufficient lighting. New lighting has reduced this and also reduced the amount of powder being sprayed. Jigs are changed frequently to maintain earthing. They have found a new supplier that can supply hooks cheaper than the cost of cleaning the old ones.
- ***Cleaning.*** Cleaning methods were improved to reduce cross contamination. This included educating operators to work better to avoid getting powder on themselves and transferring this to other jobs.
- ***Reduced maintenance costs.*** The maintenance operator was sent on a course (cost £150) enabling him to service the equipment. This has saved the company £500 to £1000 per month for each spray unit.

These improvements, particularly the change in culture and training, have reduced exposure to dust, improved quality, and reduced plant running costs. Workers have a greater understanding of their work through training and think more about why they need to follow the plant operating procedures.

Provided by Peateys Coatings, Leeds.

Note: cyclones remove the larger particulate and therefore increase the life and efficiency of the primary filtration, as well as reclaiming the used powder. Optimising inward air velocity should be achieved through matching fan size to the booth volume and openings. Peateys Coatings found the above to be the best solution for their needs, but you may decide that this not the best option for your particular circumstances.

Case study 1

Case study 1

How do I know if there is a problem?

17 There are a number of steps in determining whether your workers are over-exposed to coating powders. These are described below. It is also worth noting that powder coating plant technology and the common problems that can result in increased operator exposure are described in this guidance. The sections on best practice will also assist in determining whether you can improve the methods you use to reduce your employees' exposure. Employers following the principles of this guidance on best practice are likely to have made major steps towards reducing exposure.

18 Listed below are factors that may suggest that your plant is not operating at its optimum. These may in turn suggest that workers are being exposed to too much dust when working with coating powders. For example, workers should not be covered in coating powder at the end of the spraying activity. If your workers are covered in coating powder (see Figure 1 on page 14) then it is likely that they are being over-exposed. This is discussed further in the following sections.

Is the problem obvious?

- Workers are complaining about ill health or the dust levels.
- Workers' hands and faces are covered in powder after spraying.
- Too much powder is being wasted.
- Regular complaints from customers about quality.
- The level of re-coating is high.
- Excess powder is being deposited on the floor and application equipment.

Find out about the coating powder

19 Find out about the coating powders used by your company. The following information should be provided in the supplier's Safety Data Sheet and some details may also be found on the label on the box containing the powder. If the information is not provided or is insufficient then contact your supplier for help.

- Immediate or long-term health effects.

- Whether there are OELs for any of the constituents of the coating powders your company uses. An explanation of OELs can be found in paragraph 16. If there is no limit you need to reduce exposure to a level which will protect your workers' health by following the principles outlined in this booklet to reduce exposure to as low a level as is reasonably practicable.

- Coating powders are dusty materials, and under the COSHH Regulations 1999,[2] one of the definitions of a 'substance hazardous to health' includes *'dust of any kind, except dust assigned an OES or MEL (where different values apply), when present at a concentration in air equal to or greater than 10mgm^{-3} of total inhalable dust, or 4mgm^{-3} of respirable dust, as a time weighted average over an 8 hour period'.* These values would therefore apply where there was no specific OEL or other degree of concern to warrant control to lower levels. Employers should check whether a specific OEL is listed in EH40 for any of the component substances present in the coating powders used, and ensure compliance to that OEL. 'Total inhalable dust' approximates to the fraction of airborne material which enters the nose and mouth during breathing, and is therefore available for deposition anywhere in the respiratory tract. 'Respirable dust' approximates to the fraction of airborne material which penetrates to the gas exchange region of the lung.

- Whether components are absorbed through the skin. If so, biological monitoring (eg urine tests) may provide useful additional information. Biological guidance values and a detailed guide about biological monitoring are published by HSE, *Biological monitoring in the workplace: A guide to its practical application.*[4] If biological monitoring of workers is helpful then you will need a nurse or doctor to help you with the monitoring and to interpret the results. HSE's Employment Medical Advisory Service can give advice, including using the services of an Appointed Doctor.

Check list – hazards

	Y	N
Do I know the health effects?		
Are there any exposure limits for the constituents of my coating powders?		
Do I need to carry out biological monitoring?		

How does dust become airborne?

20 The factors that can cause dust to become airborne and stay airborne include:

- the 'dustiness' of the material (ie how easily does it become airborne - for example flour dust will blow about more than salt, because it is finer and lighter);
- the nature of the process (ie spraying powder will actively make the dust airborne);
- the extent to which the material is disturbed as a result of the jobs being carried out (for example, brushing will generate more dust than using a vacuum cleaner); and
- sources of air movement (fine dust will follow air movements and will only settle on surfaces slowly).

21 Therefore when using coating powders, which are sprayed forming a cloud of dust, there is the potential for high exposures, where adequate controls are not in place. The cloud of sprayed coating powder will follow air movements, and any fine particles that are not extracted by ventilation will settle very slowly. Workers may therefore be exposed to the initial cloud of dust and then to dust particles that are suspended in the air.

Find out about the process and the exposure

22 **Watch** the process and talk to the operators and safety representatives - you cannot assess the workers' exposure from your desk. **Don't forget** to consider activities such as filling hoppers, maintenance, work in confined spaces and cleaning of equipment. It is often the non-routine or end of the shift tasks that give rise to the highest exposures.

23 Think about how the dust is being generated and how the workers come into contact with it. Consider how often and for how long workers are exposed, and how much coating powder they are using. Understanding the factors that contribute to the workers' exposure will help you to assess and decide how to control it. Different workers may do the same job differently. For example, one may pour the coating powder gently into the hopper, whereas a colleague may tip it quickly generating more airborne dust. However, best practice would be to use a scoop to carefully transfer the powder.

24 Your workers' exposure to dust from coating powders results from any activity that causes the powder to become airborne. During the use of coating powders the activities that give rise to occupational exposure are:

- transferring the coating powder from the supplier's box to the hopper delivery system;
- fluidising the powder with air;
- spraying the coating powder;
- booth cleaning / colour changes;
- powder reclaim / sieving;
- disposal of empty product boxes;
- general plant cleaning; and
- maintenance.

25 The significance of occupational exposure to dust during the above activities depends on how the work is carried out and the controls in place.

26 If you do not know whether your workers are being over-exposed (please read this guidance in full before making this decision), then you may need to measure the workers' exposure. You can either do this yourself or employ the services of a consultant (see Appendix 1). However, if it is obvious that you are not adequately controlling exposure then first make the improvements before carrying out any measurements. If you are sure that you are adequately controlling occupational exposure then you may not need to measure their exposure, although for most powder coating operations air sampling is likely to be needed.

First make the improvements to your plant and powder coating operation before deciding what air sampling you may need to do.

27 Where exposure is at or above 10 mgm^{-3} of total inhalable dust (see the definition in paragraph 19) over a working shift, it may not be visibly obvious that exposure is high. Some companies perceive the concentration of 10 mgm^{-3} 8-hour TWA to be a 'thick fog', which is more typical of the explosive dust concentration of 10 grams m^{-3}. The plant may appear to be clean and tidy and the booth operating efficiently. However, the operator may still be over-exposed if spraying and cleaning practices are poor.

It is the dust concentration inside the booth that matters and whether the operator is standing inside that cloud of dust, and not just the cleanliness of the exterior of the plant.

28 The following table shows the results of measurements taken by HSE in 1994 to determine the degree of occupational exposure to the total dust in the air that workers were exposed to. The table shows the extent of exposure that the operator may experience averaged over a full working shift. Operators may receive exposures higher than these values during particular tasks, such as cleaning, followed by periods of lower exposure. These results represent exposures where the company was failing to adequately control occupational exposure. However, it should be noted that HSE did not specifically visit companies that were known to have high exposures. They were selected as typical powder coaters.

HSE survey of powder coaters 1994

Task	Average Exposures	Maximum Exposures	Extent over 10 mgm^{-3} (see paragraph 19)
manual spraying	11 mgm^{-3}	53 mgm^{-3}	5 times greater
automated spraying	21 mgm^{-3}	79 mgm^{-3}	8 times greater
loading of powders	4.7 mgm^{-3}	17 mgm^{-3}	2 times greater
cleaning activities	59 mgm^{-3}	131 mgm^{-3}	13 times greater

Note 1: exposures were all generally high.
Note 2: exposures were high for automated spraying where manual touch up was also carried out.

29 Operators were seen to lean into booths to spray the articles, and cleaning practices were very poor, including the extensive use of compressed air. Your workers may have similar high exposures if the plant is not set up correctly, or if they follow poor working practices such as leaning into the booths to spray articles or use compressed air extensively.

When a worker is over-exposed to coating powder then it probably means that your company is wasting coating powder.

30 It is also important that you consider skin exposure as well as what the workers may be breathing in. Workers may routinely come into contact with the powder from the box or from contaminated surfaces. Workers' hands and faces may become coated in powder during spraying and cleaning activities, where exposure has not been adequately controlled. Some powders can cause direct effects on the skin (see paragraph 12).

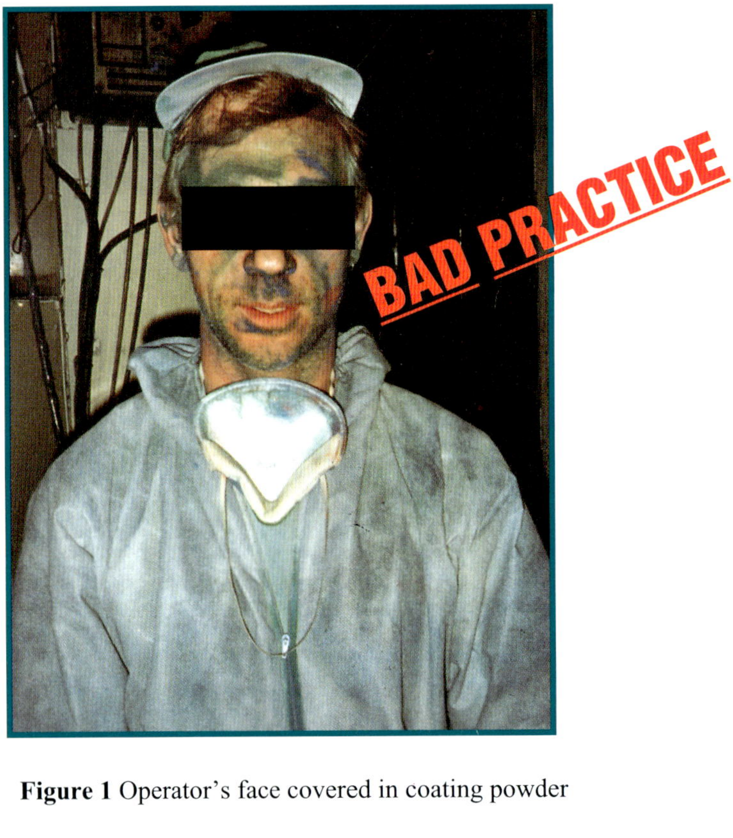

Figure 1 Operator's face covered in coating powder

How do I know if the exposure is high?

31 If you have measured exposure and the substance has an exposure limit then you can compare your results to this value. Since coating powders are mixtures of various components you should ensure that exposure to each component is controlled. Where an OEL is assigned to a specific substance, for example TGIC, then exposure to that substance should be controlled below its limit. However, in many cases there will not be an exposure limit and you will need to ensure that exposure over the working shift is controlled to at least below 10 mgm^{-3} total inhalable dust. Where there is evidence to suggest that control to lower levels is needed to protect workers' health, then you should reduce exposure further.

32 Appendix 2 provides guidance reproduced from HSE Engineering NIG sheet no 15 on the use of banding limits for assessing exposure to TGIC through the measurement of total inhalable dust.

33 You should also check your health records and talk to staff and safety representatives to find out if there have been any cases of ill health that can be attributed to exposure to coating powders. This may include workers complaining of sore throats or itchy hands and faces.

34 While you should always consider the above approach to find out whether you have a problem with controlling exposure, you should also consider whether you are following best practice. You could talk to other employers in the industry, your coating powder or equipment supplier, or your trade association (see Appendix 1). The following sections consider each aspect of the use of coating powders for electrostatic spraying, how exposure arises, how this can be reduced and what new technology is available. They consider the list of activities on paragraph 24 and how changing the way you work can reduce the amount of powder from becoming airborne, reducing both the workers' exposure and wastage of the coating powder.

Check list – exposure

	Y	N
Is there too much powder in the air that workers may breathe in? (see paragraph 26.)		
Do they get a lot of powder on their hands and faces?		
Have there been any cases of ill health?		

35 You should now consider what action you need to take to reduce exposure.

36 Controlling exposure within your powder coating plant will probably involve a package of measures that bring an overall benefit. Many of the aspects in the following sections are related because changing one can affect another. For example, when jigs are not properly earthed, the attraction of the powder to the article will be reduced, which may result in operators adopting poor spraying practices to achieve the desired finish.

Managing exposure to coating powders

37 The following sections guide you through the powder coating process and what needs to be considered to achieve reductions in exposure.

Substitution

38 You should first consider the range of coating powders that your company uses and whether any can be substituted for less hazardous alternatives.

You may have been using some powders for years that were originally selected simply because the colour was right. There may now be alternatives available that present a lower overall risk to health, safety and the environment.

39 If, for example, you are using a TGIC coating powder you should check to see if it is really necessary. If the object to be coated is for indoor use, a TGIC polyester powder may not be required. Where polyester coatings are essential, alternative curing agents to TGIC may be available. The ability of the alternative to do the job, and the health risks that may be associated with it, should be properly assessed. Ask your current supplier, or other suppliers, for information about alternative less toxic coating powders, but take care not to substitute one health risk for an alternative with a greater overall health, safety and environmental risk. A material that is reported as 'safe' may simply mean that less is known about it. Your workers' health may still be at risk even if you are not using coating powders containing TGIC, if working practices are poor. Further details are given in *Seven steps to successful substitution of hazardous substances.*[5]

Check list – substitution

	Y	N
Can we use an alternative?		
Does the alternative present a lower overall risk?		
Will it still do the job as effectively?		

Using coating powders correctly

40 The following sections look at how coating powders are used, how exposure arises and where common problems exist. It is important to understand that the way the booth is set up and used can affect the worker's exposure. Controlling exposure is not just about the efficiency of ventilation systems, but the overall set-up of the plant, its maintenance and the operator's training and awareness. The latter is particularly important. Without the adequate training of your operators it is unlikely that you will achieve significant reductions in exposure. In addition to reducing exposure, these factors can also reduce your process running costs and improve quality.

Powder delivery systems

41 Delivering the coating powder to the gun ready to use is the first point of the process at which workers may be exposed. There are two ways in which this process is carried out:

- transferring the powder from the supplier's box to the hopper on the powder delivery system (Figure 2). Once the powder is fluidised it is ready for supply to the gun; and

- direct-box-feed systems (Figure 3). With these systems the supplier's box is vibrated to fluidise the powder ready for delivery to the gun. Powder transfer from the box to the hopper is not necessary. There are other systems which avoid the need for transferring powder to a hopper, for example, big bags, durabins and kegs.

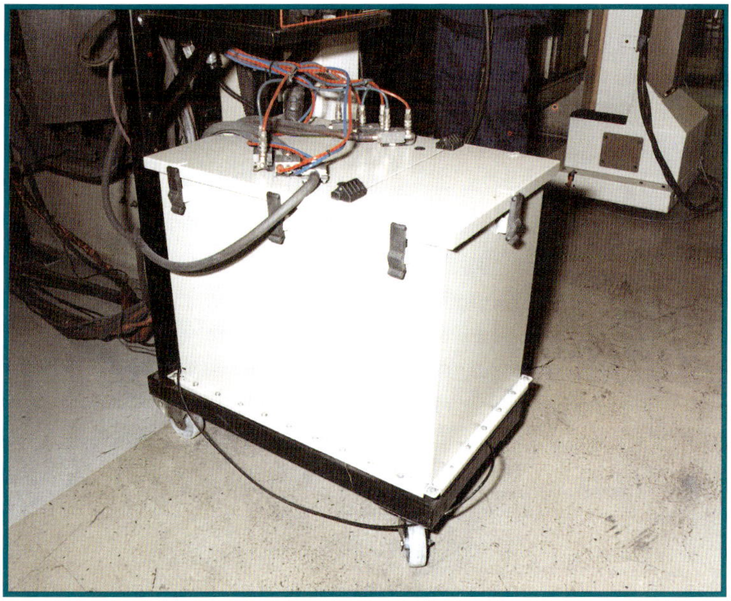

Figure 2 Coating powder hopper

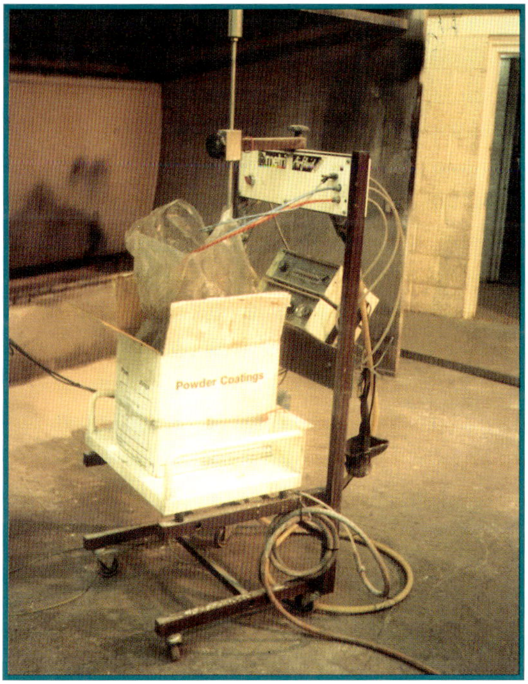

Figure 3 Direct-box-feed system

42 Occupational exposures are likely to be higher with the traditional hopper system in Figure 2, because the worker has to pour the powder from the supplier's box into the hopper. It is therefore likely that spillages will occur and powder will become airborne as it is poured. The extent of this exposure will depend on how carefully the worker pours the powder. Direct-box-feed systems will overcome this as the gun delivery tube is simply lowered into the supplier's box. It is, however, possible to adequately control exposure with traditional hopper systems if sufficient care is taken. Where exposure is found to be high then local exhaust ventilation may be needed. Any such investment would need to be balanced against the use of direct-box-feed systems. There may also be specific requirements for the coating powder in use that dictates one system over another. If local exhaust ventilation were used then it would need to semi-enclose the hopper and provide an inward air flow sufficient to prevent the escape of dust, typically 0.5 to 1.0 metres per second. It should also be noted that with both systems coating powders are being fluidised with air with the potential for generating airborne dust. Therefore consideration should also be given to this source of exposure, for example, venting into the booth to reduce worker exposure.

43 Care needs to be taken with both methods during the disposal of the supplier's coating powder boxes. Whenever a container is squashed for disposal the air inside will be forced out carrying with it any residual powder, which may be breathed in by the worker. If compressed air is used for cleaning the delivery system, this increases the potential for high exposures and should be avoided.

Coating powder application - the booth system

44 The powder booth provides an environment for the coating powder to be sprayed onto articles and excess powder to be reclaimed or sent to waste. The released powder from the gun is contained within this environment, with little or no release into the workplace if the booth is set up correctly. The operator's exposure is dependent not only on the booth efficiency, but how he / she interacts with the process, as follows:

- the extent to which manual spraying is carried out (including 'touch-up' for automated spray lines);
- where manual spraying is carried out, the way in which the operator sprays the coating powder, and the extent to which he / she leans into the booth;
- the nature and size of the article being sprayed, and whether spraying is wrongly carried out outside the confines of the booth; and
- the transfer efficiency of the coating powder, which results from how the equipment is set up and the powder characteristics.

45 These points are discussed in detail in the following sections. In this section the focus is on the correct setting up of the spray booth. There are many different spray booths available for the application of coating powders, which generally fall into two categories. These are:

- conveyorised tunnel booths (manual and automatic) - see Figure 4; and
- simple open fronted manual spray booths (traditionally used for wet paints) - see Figure 5.

Figure 4 Conveyorised spray booth

Figure 5 Traditional open fronted spray booth

46 Some booths are fitted with powder reclaim, where more than 90% of powder is recovered for re-use, with the remainder going to waste. Whatever booth type is used, they work on the basis of preventing over-sprayed powder from escaping into the workroom atmosphere, and where powder reclaim is carried out from being wasted.

47 To achieve adequate control, the rate of velocity of the air entering the booth should be greater than the velocity of the dust being released towards the booth openings. Generally, air entering the booth should be between 0.5 and 1.0 metres per second at the booth openings. This can be checked using an anemometer (a small device to measure air movement), or judged by using smoke tubes or a Tyndall beam. Smoke tubes allow you to release smoke at the openings to determine directions of airflow and make a judgement as to the efficiency of the ventilation system. A Tyndall beam is a powerful light source (a powerful torch may be sufficient) that can be positioned to enable the fine dust to be seen and determine whether it is being controlled (similar to the effect sunlight has shining into a room through a window). If you do not have access to any of the equipment described above, you may be able to judge whether the booth is adequate by simply watching the powder as it is carried by the air movement. Whatever method you use you are checking that the booth is pulling the powder away from the worker as he / she carries out the spraying. Simple checks such as those described above allow for the early detection of deterioration of the plant efficiency. Typically these checks may be carried out daily or weekly. It is also important that the dust is always drawn away from the workers breathing zone and not into it (see paragraph 50).

48 The COSHH Regulations[2] also require that local exhaust ventilation (LEV) systems are thoroughly examined and tested at least once every 14 months, except with certain processes where it must be carried out more frequently. This includes the above aspects and others such as the velocity of the air moving through the ducting (the transport velocity), which needs to be high enough to prevent the dust settling and causing blockages in the ducting. The condition of the filtering equipment should also be checked regularly. HSE produce guidance on this subject, *Maintenance, examination and testing of local exhaust ventilation HSG 54*.[6]

49 Care should be taken to avoid using too high velocities to extract air (see paragraph 47) as this could affect the coating powder transfer efficiency as the charged cloud of powder is drawn away from the work-piece. This would result in a need to spray more powder to achieve the desired finish.

50 Open fronted booths should be large enough to allow the article being sprayed to be positioned in the confines of the booth. Articles should not be sprayed outside the booth. The air outside the front of the booth will only move towards the booth slowly and will be subject to turbulence from the worker's body. The operator should not stand between the article being sprayed and the booth extraction whilst spraying. This allows dust-laden air to be drawn across the operator's breathing zone. Figure 6 demonstrates these two poor practices. Where necessary, the article should be repositioned using a turntable or jig to allow each side to be sprayed, rather than the sprayer walking around it.

Figure 6 Operator standing between booth extraction and article being sprayed and note article extends outside the front of the booth, where air extraction rate is lower - see paragraph 50

Case study 2

Pearlvale are a small trade coater who use two small open-fronted booths and one small conveyorised spray booth.

The company were concerned that the exhaust ventilation on the open-fronted spray booths was not efficiently removing airborne powder during spraying. They were aware that the velocity of the air entering the front of the booths was too low. They realised that if they reduced the size of the openings the air would move faster through the smaller openings.

They had small panels fabricated that could be hinged to the sides of the openings, which could be left open or closed depending on the size of the piece being sprayed. The excess powder is now more efficiently removed during spraying and the workers' exposure has therefore been reduced at minimal cost.

Provided by Pearlvale Limited, Hall Green, Birmingham.

Case study 2

Case study 2

51 Since the powder does not leave the end of the spray gun at high velocity, it is possible that the velocity of the air entering your system will already be high enough to prevent releases into the workplace, providing it is between 0.5 and 1.0 metres per second at the booth openings. If needed, the air velocity can be increased by reducing the size of openings where possible, for example, the exit / entry for articles and the openings for spraying (see Case Study 2).

- ***Inward airflows into spray booth openings should be between 0.5 and 1.0 metres per second.***
- ***Articles should always be sprayed within the confines of the booth.***
- ***The operator should not stand between the article being sprayed and the booth extraction.***

52 Even if the booth is operating efficiently, exposure may still be high depending how the worker carries out the spraying activity (see paragraphs 64 to 70).

Check list – booth

	Y	N
Airflows optimised (0.5 to 1.0 m/s)?		
Booth large enough for articles?		
Openings minimised?		
Examined and tested?		

Coating powder booth developments

There have been various technological developments in spray booth design. Coating booths are now available that include the following developments:

- booth construction materials that repel charged powder and therefore reduce surface deposition. These reduce waste and the extent of cleaning that is needed;
- systems with multicyclone filtering arrangements for multi-colour operations;
- automated guns that can be programmed to recognise different shaped work-pieces and co-ordinate gun movements; and
- coating powder booths with integral cleaning systems (see 'Booth cleaning developments' below paragraph 76).

The inclusion of an item above does not mean that it is endorsed by HSE. Each system should be judged on its merits for the particular plant for which it is intended. For further information on these systems you should contact your equipment supplier or trade association (see Appendix 1).

Coating powder application - other plant set-up factors

53 There are other factors that can increase the efficiency of your powder coating plant and maximise powder delivery. These include:

- Jigs can become coated in over-sprayed powder and lose their earthing potential. This results in a decrease in the transfer efficiency of the powder, which can result in higher exposures and increased material wastage. This may also result in an increase in the amount of re-coat work that is carried out as a result of poor quality finishes. The jigs holding the articles should therefore be cleaned on a regular basis to maintain the transfer efficiency of the powder. The paint can be either burnt off or chemically stripped. This is a specialised operation and you should seek advice about these methods or contract out the work to an appropriate company. Some companies have found that it is more cost effective to replace the jigs than clean them.
- The shape of the jig hooks can affect the earthing of the article. 'S' shaped hooks with smooth curved grooves allow better contact between hook and article. For sharp angled hooks, paint will build up quicker at the contact point of the hook and article. These will therefore require more frequent cleaning or replacement.

- Loading the maximum number of articles (particularly for small parts) onto the jigs, without overcrowding them, is also very important when maximising the efficiency of the process. This is particularly the case with automated guns but also important with manual spraying. As the article being sprayed passes in front of the guns, exposing the greatest possible surface area to the sprayed powder will increase the transfer efficiency, reduce the time for coating runs and reduce occupational exposure.

Check list – plant

	Y	N
Jig loading maximised?		
Article earthing maximised?		

Coating powder application - spray equipment

54 There are two types of powder coating spray guns, namely:

- corona charging; and
- tribo charging.

55 With corona charging, a high voltage is applied to a corona charging point, a needle, near the air-blown powder. Negatively-charged ions are created by the corona discharge from air molecules and become attached to powder particles being discharged from the spray gun. Discharged ions and charged powder particles are pushed down the lines of the force created by the corona discharge to the nearest electrical earth, the work-piece.

56 In tribo charging, the powder particles are passed over a plastic surface, eg polytetra fluoro ethylene (PTFE), in the spray gun. Friction rubs electrons off the powder particles, giving them a positive charge. To ensure good charging, the powder passages in the spray gun are narrow and maze like. The gun is earthed to prevent the build-up of negative charge. The positively charged powder particles are then blown towards, and attracted to, the earthed work-piece.

57 There are electrostatic effects that can affect the ability of the powder to coat the article. The extent of these effects will depend on the choice of gun and how it is set up. These electrostatic effects include:

Faraday cage effect - Free ions that are not attracted to the work piece can build up in recesses of the earthed work piece, creating pockets of negative charge, which repel the charged powder particles. This results in a difficulty in applying coating into corners and enclosed areas as the particles take the path of least resistance and do not deposit in these areas. The Faraday cage effect is more significant with corona charging than tribo guns, although it is understood that the latter are not suitable for the application of all powders. The Faraday cage effect is also less significant when applying to relatively flat surfaces and when the voltage is reduced.

Back ionisation - The corona process of electrostatic charging is the most commonly used and allows fast transfer and efficient, easy application. However, it often introduces film defects associated with the variation in powder charging levels and free ion production in the air stream. 'Free ions' are charged particles of air. They are similar in polarity to the powder so will repel adjacent charged particles, whether air or powder. These particles can disrupt the applied powder film causing defects, which often show as pockmarks and uneven coatings and which can be unacceptable. This is phenomenon is referred to as 'back ionisation'.

Sometimes corona charging guns are fitted with free ion collecting devices that can avoid some of these problems. These are earthed needles, placed external to the gun and adjacent to the spray nozzle, which work by stripping the free ions and discharging them to earth in the area between the gun and the component. These devices will often aid penetration into enclosed areas as well as giving smoother and more even coatings. However, they will often lower the transfer efficiency of the powder coating process.

58 As a result of affecting the ability of the powder to coat the article, these effects can influence the way the operator sprays the powder, and therefore his or her exposure to coating powders. They should be minimised where possible by proper control of gun settings.

59 The factors that determine the required powder delivery rate, the airflow and the voltage necessary are:

- the shape and size of the article;
- the coating powder characteristics;
- the required finish; and
- the above charging effects.

60 Generally it should not be left to the discretion of the operator to choose the gun setting. These settings will vary from job to job depending on the factors in paragraph 59 and should be predetermined. It is good practice to use job specification sheets that inform the operator what setting to use for that particular job. Failure to control these settings can result in high powder usage, poor powder attraction, reduction in quality and an increase in re-coat work. For example:

- Increasing the powder delivery rate beyond that needed increases over-spray of powder, product waste and occupational exposure. It can also affect product quality.

- Increasing the voltage charge to the powder will increase charging of the powder, which will in turn increase the potential for ionisation effects, such as the Faraday cage. The operator may then turn up the powder delivery rate or attempt to move in closer to overcome this. This again results in product waste and higher exposures.

61 It is important that once the type of electrostatic charging system has been selected, the appropriate settings are used to apply the powder. These settings will vary depending on the powder being used, the article and the desired finish. For general applications the voltage should typically be 75kv.

Poor control of gun settings, with too high powder delivery rate can result in increased exposure and product wastage.

62 There are many different guns available, which make different claims about the extent to which the above ionisation effects can be reduced. The merits of any proposed capital expenditure should be carefully considered against the likely benefits and the extent to which you can reduce exposure through good practice.

Automated spraying can reduce occupational exposure.

63 The coating powder can be applied through either automated or manual spray equipment. Automated spraying can reduce occupational exposure. You should consider this approach to control, balancing any capital expenditure against reductions that can be achieved through good manual spraying practice. Although some of the above difficulties are removed when using automated spray guns, there may still be a need for manual 'touch up' to coat areas missed by the automated guns. Uncoated areas are usually as a result of a failure of the automated guns to apply coating in recesses of the article where the Faraday cage effect occurs. Automated guns reciprocate horizontally or vertically (most common), or 'waggle' vertically and do not always have the ability to overcome these effects. It is, however, possible to place a second set of automated guns before or after the main coating guns to apply coating powder at a

lower voltage to reduce the effect and achieve coating of these recesses. Similarly, when using manual guns to 'touch up' these areas, the voltage can again be reduced. However, manual 'touch up' should be kept to a minimum as it can also result in high exposures.

Check list –spray equipment

	Y	N
Automated spraying considered?		
Gun selection checked?		
Gun settings optimised (air, voltage, powder)? *(general applications, voltage should be 75kv)*		
Ionisation effects minimised?		

Coating powder application - spraying technique

64 If all the guidelines provided in this booklet so far have been followed then the booth should be running at optimum efficiency and the guns should have been set up correctly. All these benefits are nullified if the operator adopts poor working practices. It is often the way the operator carries out the spraying that results in the high exposures. Paragraph 28 shows the high exposures found during the application of coating powders, which have resulted in many cases from poor spraying practices.

65 These poor spraying practices usually result from poor training (see paragraphs 93 to 98). The lack of training results in workers believing that they have to point the gun close to each surface of the article and rock it back and forth, failing to understand the principles of this coating technique. Typically operators have been found to:

- lean too far into the booth to enable the gun to be pointed at each surface and become enveloped in the cloud of charged powder;
- position the gun closer than needed to the article; and
- where two operators worked in the same booth, spray from opposite sides of a booth with the guns pointing at each other's breathing zone, increasing their work colleague's exposure.

Case study 3

Case study 3

Baxi Heating, manufacturers of central heating boilers and gas fires, have an in-house powder coating plant. They operate three automated coating lines. Their concern to ensure high quality finishes led them to strictly control plant operational factors. To achieve this high standard they carried out the following:

- Spray guns. Strict operating procedures are provided for each job. Gun settings are detailed on the job specifications to optimise powder usage and maximise quality. These are determined for each new job through trial runs.

- Jigs are cleaned after every three runs to maintain good earthing of the articles.

- Contamination. Each colour has its own fluidised bed to avoid cross contamination.

- Loading of jigs is designed to ensure the maximum number of articles are sprayed for each pass of the guns, reducing powder usage.

- Cleaning. Scrapers are used for cleaning, with compressed air only used for the very final clean. Operators wear power assisted respiratory protective equipment during cleaning/colour changes.

Optimising plant set-up, for example gun settings and jig loading, reduces occupational exposure, reduces plant operating costs through reduced material use, and improves quality. Re-coat is also reduced, which again reduces plant operating costs. For Baxi, this has allowed them to compete better for trade coat work where high standards of coating are demanded.

Provided by Baxi Heating, Bamber Bridge, Preston.

Case study 3

66 Good and bad spraying practices are shown in the figures below.

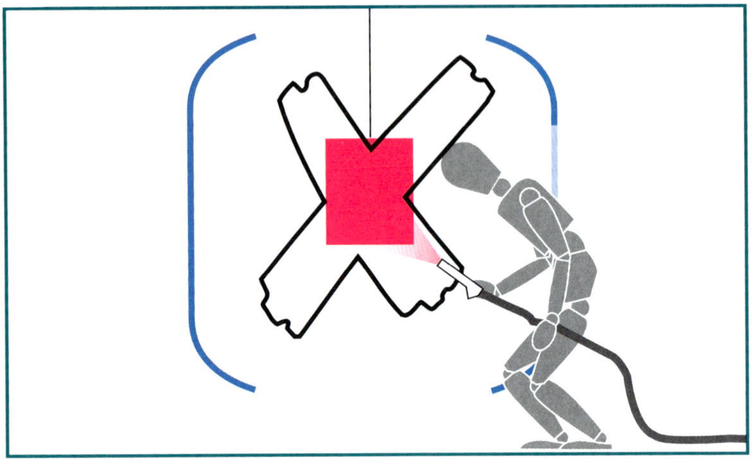

Figure 7 Worker leaning too far inside the booth

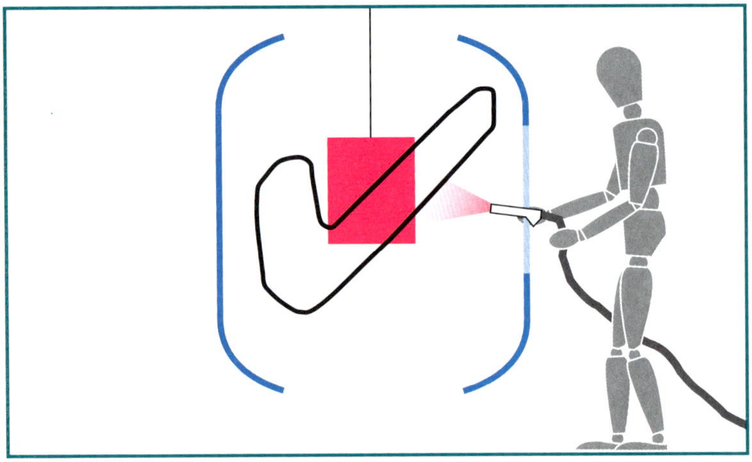

Figure 8 Worker standing outside the booth to spray

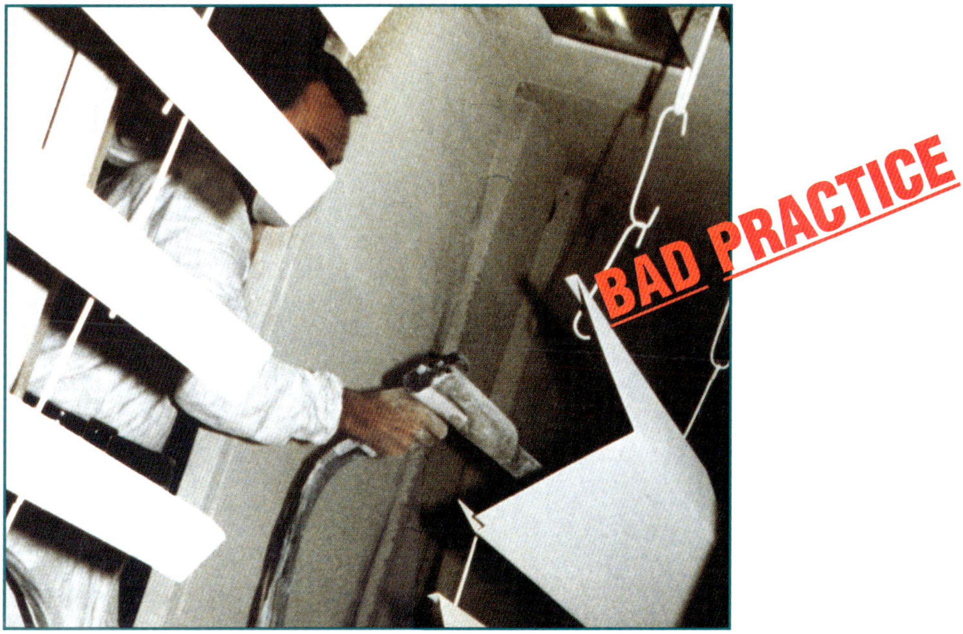

Figure 9 Operator leaning into a booth to spray

Figure 10 Operator stood outside the booth to spray.

Note to Figure 10: Whilst control should be achieved without the operator having to wear RPE, this may not always be possible even with good working practices (see paragraphs 83 to 91).

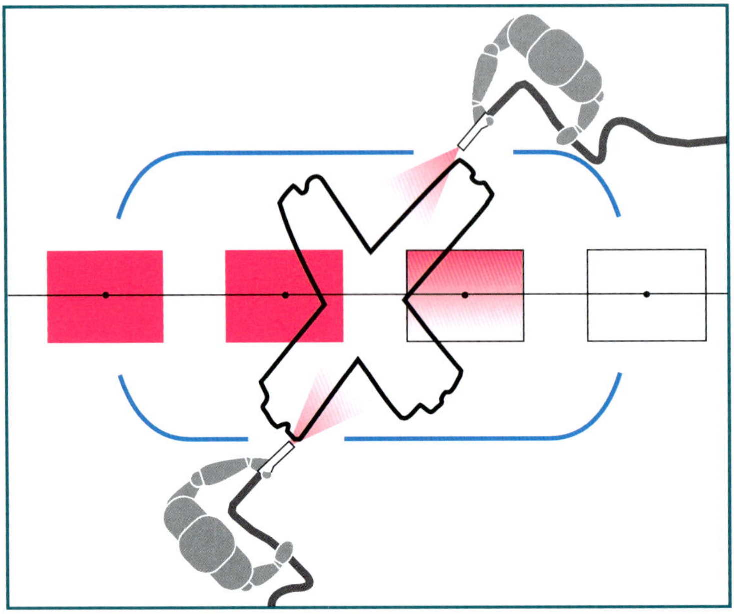

Figure 11 Overhead view of operators spraying towards each other

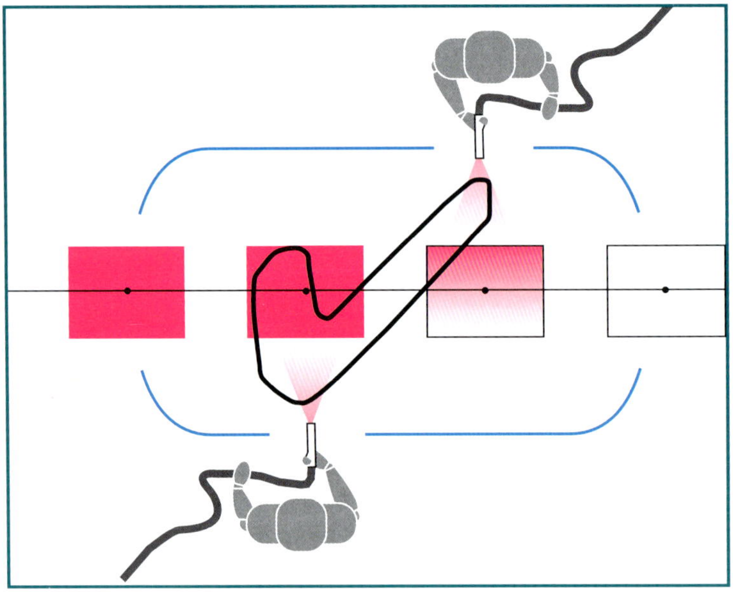

Figure 12 Overhead view of operators spraying straight ahead to avoid spraying towards each other

67 Poor spraying practices can be exaggerated by other process factors that cause the worker to compensate for poor powder transfer efficiency (as described in earlier sections). These may result in the worker spraying more powder than needed, leaning into the booth and moving the gun very close into crevices to overcome poor attraction of the powder to the article. These factors include high powder delivery rates, resulting in excessive over-spray; high voltages which can over-charge the cloud of powder and reduce its attraction to the article; and poor earthing of articles reducing the attraction of the powder to the article.

68 These factors may be due to poor training and awareness, and insufficient supervision of plant set-up.

69 With automated systems there may still be a need for manual 'touch up' to finish those parts of the article missed by the automated guns (see paragraph 63). Where automated guns are in use, manual 'touch up' work should be kept to a minimum and avoided if possible. Flat and less intricate objects are unlikely to need manual 'touch up' spraying. Where this practice is carried out poor spraying practices may be evident.

70 Another problem with automated guns is where the upper angle of the reciprocating guns results in powder being sprayed out of the top of the booth (see Figures 13 and 14). Where the guns reciprocate too low powder can be sprayed direct into the extraction system and wasted.

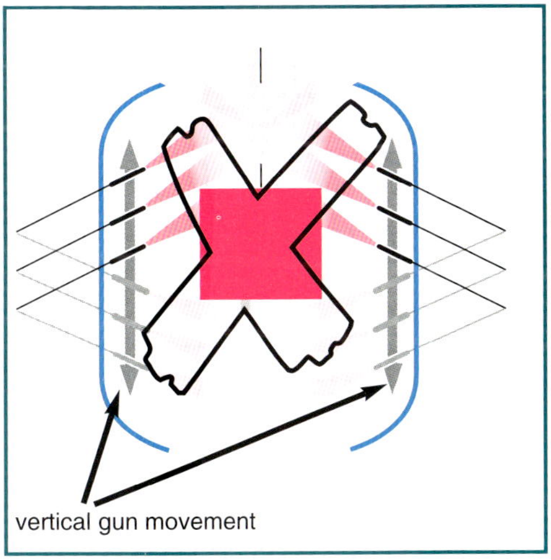

Figure 13 Angle of gun reciprocation too high or too low

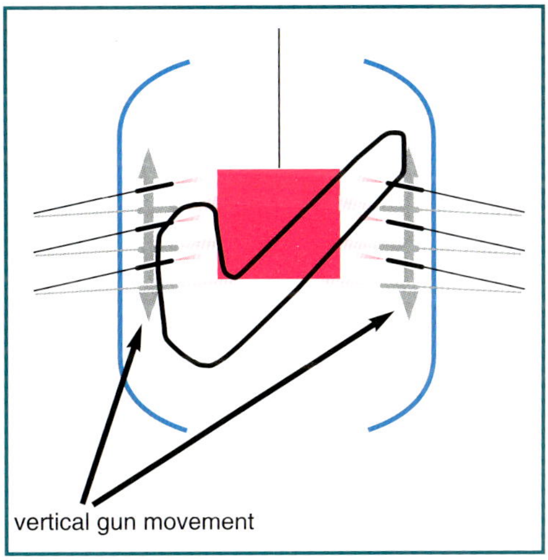

Figure 14 Angle of gun reciprocation prevents release from booth

Check list – spraying technique

	Y	N
Operators not leaning into the booth?		
Operators not spraying towards each other?		
Automated guns positioned correctly?		
Operator not changing settings incorrectly?		

Cleaning methods

Figure 15 Operator brushing the inside of a spray booth

71 Cleaning of the interior of the powder coating plant is fundamental to the efficient running of the process, and high standards of cleanliness avoid contamination of material. The need for a high standard of cleanliness of the interior of the plant and for this to be carried out quickly can result in very high exposure to coating powders. Very high exposures result from the poor methods often used to clean the inside of the booth. These include:

- workers standing inside the booth whilst cleaning;
- the use of brushes; and
- the extensive use of compressed air.

72 The use of brushes and compressed air create very high exposures, to the extent that the operator can be enveloped in a cloud of dust. Brushes and compressed air are also an inefficient way of removing the powder as they will first redistribute the material around the booth.

73 You should use a cleaning regime that minimises the worker's exposure. The following points should be considered:

- leave the ventilation system running during cleaning;
- the operator **should not be standing inside the booth.** Long handle tools can be used to avoid this where possible;
- brushes should never be used. Rubber scrapers that allow the material to be pushed / scraped off the inner surface should be used;
- as much powder as possible should be removed using the scraper before using compressed air;
- industrial vacuum cleaning equipment should be used where possible;
- impregnated wipes can help to remove damp residues that vacuum cleaners cannot remove;
- the use of compressed air should be kept to a minimum and only for the very final clean and reversible air guns used if possible; and
- the velocity of the compressed air should be kept to the minimum required to remove residual powder from awkward areas.

74 Even if you follow the above guidelines it is likely that the operator will need to wear respiratory protective equipment (see paragraphs 83 to 91). Many of the other factors in this guidance can also affect the amount of cleaning needed and therefore the level of exposure. Optimising booth and gun settings can reduce the amount of over-spray and therefore the extent of powder deposition on internal booth surfaces.

75 There are new booth developments that incorporate automatic cleaning (see 'Booth cleaning developments' below paragraph 76). These can reduce cleaning times and increase plant efficiency. However, the merits of any proposed capital expenditure should be carefully considered against the likely benefits and the extent to which you can reduce exposure through good practice.

76 The methods used for carrying out general plant cleaning should follow the same principles of minimising dust generation. For general plant cleaning, industrial vacuum cleaners should be used and compressed air only used when necessary, for example, controlled use for small crevices in plant. Reversible air guns may also be considered as an alternative to vacuum cleaners.

Check list – cleaning

	Y	N
Brushes not used?		
Scrapers used?		
Vacuum cleaning equipment used?		
Compressed air kept to a minimum?		
Operator not entering the booth to clean?		

Booth cleaning developments

- *Automated water cleaning techniques*

  Booths that can be totally enclosed to allow washing of the internal surfaces and then drying on a preset programme, reducing cleaning time from 60 minutes to about 20 minutes.

- *Automated air cleaning techniques*

  Booths that can be closed to allow fixed air jets to blow the internal surfaces clean, with only minimal use of compressed air lances to do a manual final clean. These can also reduce cleaning times to about 20 minutes.

- *Automated wall scrapers*

  Fixed scrapers that mechanically run the length of internal surfaces to remove deposited powder, reducing clean times.

- *Single use booths*

  Single use booths are also available where each colour has its own collapsible fabric booth. Colour changes simply require the operator to change the collapsible booth in use under extraction and replace it with the booth for the next colour.

The above are provided for information about new technological developments. The inclusion of an item above does not mean that it is endorsed by HSE. Each system should be judged on its merits for the particular plant for which it is intended. For further information on these systems you should contact your equipment supplier or trade association (see Appendix 1).

Minimising process costs

77 The previous sections demonstrate how reducing worker exposure can reduce process running costs. This section summarises these points. Possibly the greatest savings can be made from reducing the amount of coating powder used. This can be achieved through maximising coating powder transfer efficiency. The greater the percentage of sprayed coating powder which deposits on the article the less that needs to be sprayed to achieve the desired coating. The article will also become insulated when coating has been achieved, and any further application of powder will simply increase wastage.

- ***A well operated and maintained powder coating line can increase the amount of coating powder sticking to the article from about 30% to about 70%.***

- ***Where over-sprayed powder is not reclaimed, so called 'spray-to-waste', maximising the amount of powder sticking to the article may be more important.***

78 In summary, increasing transfer efficiency is achieved through:

- powder selection;

  Selecting the optimum powder for the job that maximises the amount of powder sticking to the article with the desired finish quality. Talk to your powder supplier(s) about powder selection and optimising powder transfer efficiency.

- correct gun settings;

  1 voltage
  2 powder delivery rate
  3 air flow

- sufficient earthing of components;

  Maximising the earthing of the articles increases the attraction of the powder to the article and therefore reduces wastage.

- optimum jig loading; and

  Maximising jig loading will increase the amount of powder released from the gun reaching surface areas to be coated.

- operator training.

  Poor awareness leads to poor control of gun settings that can lead to increased powder use and increased exposure. Poor spraying practices also result in workers spraying too much powder.

79 Where quality is poor as a result of poor control of plant operating conditions, articles may need to be re-coated. This will result in increased operating costs, and further exposure to the operator.

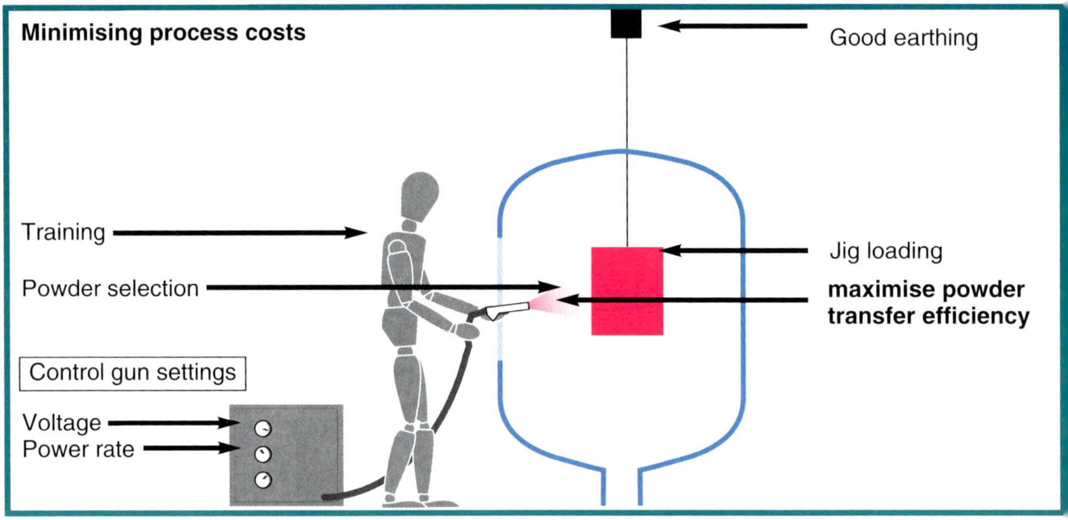

Figure 16 Factors to consider for the minimisation of process costs

Personal protective equipment (PPE)

80 You should use means other than PPE (gloves, aprons, boots, safety spectacles, respiratory protective equipment etc) to prevent or control exposure to coating powders, so far as is reasonably practicable. In other words PPE is the last line of protection.

81 Direct controls at source should always be the primary means of preventing/reducing exposure. However where it is not reasonably practicable to improve their performance to achieve adequate control, then suitable PPE including respiratory protective equipment (RPE) may be needed.

82 Suitable PPE including RPE should be available in or adjacent to the work area. You will need to provide suitable storage and changing facilities and these need to be located and designed to prevent the spread of contamination from protective clothing and equipment to personal clothing.

Respiratory protective equipment (RPE)

83 As with any PPE, the selection of the right RPE is very important to achieve adequate protection. Further practical advice is given in the HSE publication *The Selection, use and maintenance of respiratory protective equipment,* HSG53.[7] After consulting HSG53, if you are still unsure about selecting the right RPE you may require advice from a health and safety expert such as an occupational hygienist and / or suppliers of RPE.

84 RPE falls into two categories:

- Respirators – these filter coating powder dusts from the air being breathed. These are unsuitable for use in confined spaces including oxygen deficient atmospheres.

- Breathing apparatus provides clean air to the wearer (see HSG53).

85 The chosen RPE must be suitable for the circumstances in which it is to be used. This means that it must be adequate in terms of the protection it affords against the airborne concentration of the hazardous substance(s); must be capable of providing a sufficient quantity of clean air; must fit the wearer (see Figure 17); must be used in accordance with the manufacturer's instructions and must be 'CE' marked.

86 Equipment that may be used for protection against coating powders is summarised in the table on page 45. Whether or not operators need to wear RPE will depend on the coatings being used and the extent of exposure. Following the guidelines in this booklet should mean that you have taken significant steps towards reducing occupational exposure without the need for RPE. Where RPE is still needed, its selection will depend on the coating being used; other hazardous substances present in the environment; type of job; wearer's facial and medical conditions; and the environment (confined spaces, spray booths, open space etc) in which the job is undertaken. The selection of the right RPE forms part of the COSHH assessment and the criteria used for the selection should be recorded. You should consult and involve employees in the selection process. This increases the potential for correct use.

87 The table describes different standards of RPE for use against coating powders. As indicated these provide increasing levels of protection, with selection dependent on the level of exposure. Higher standards of RPE may be needed where the assessment shows that due to the nature of the coating powder being used this is warranted, for example, if it contains TGIC (see paragraph 13). It may also be that the equipment is not operating at its optimum or working practices are poor. Where this is the case you should choose RPE that offers a high standard of protection whilst you work towards reducing exposure by other means as described in this guidance.

88 The protection achieved can also vary between workers due to individual face fit differences, the closeness of shaving and the care taken when wearing it (see Figure 17). Therefore you should choose RPE that provides a greater level of protection than needed to take account of these factors. For example, if the level of exposure is twice as high as the OEL then, assuming you have taken steps to reduce this, you should issue RPE that can be used up to 10 times the OEL.

Figure 17 This operator has not shaved and therefore the level of protection has been reduced. The disposable respirator also only has a single strap which indicates that it is only a 'nuisance' dust respirator and not for toxic dusts. This type only provides minimal 'comfort', primarily for the do-it-yourself market, and is not designed to reduce occupational exposure.

89 Disposable masks are easy to use and may be comfortable to wear but they do not provide protection to the face. Properly worn full face masks and visors will provide protection to the faces and eyes as well as against inhalation exposure.

90 Filters have a limited lifetime and you should seek the advice from the manufacturer / supplier on the extent of use before they are discarded. Disposable respirators are designed to be replaced after each shift. Where pre-filters are used in the filter cartridges these should be changed on a daily basis.

91 RPE users must have received sufficient information, instructions and training as recommended in HSG53.[7] RPE used at work must be regularly maintained in accordance with the manufacturer's instruction to ensure that they remain effective. Maintenance includes cleaning, disinfection, examination, repair, testing and record keeping.

Table Standards of control showing where RPE is required (see paragraphs 86-87)

Activity	Current standard of control	Respiratory protective equipment type needed (only some types listed)	Maximum use concentration against coating powders
Hopper loading	LEV	None	-
	No LEV	Disposable with P2 filters	OEL x 10
Spraying - manual or automatic	very good equipment set up and operator practices	None	-
		Disposable with P2 filters	OEL x 10
	↓	Power assisted respirator TM2 with full face mask	OEL x 20
		Powered respirators TH2 with hood	OEL x 20
		Constant flow air line BA with visor	OEL x 40
	poor equipment set and poor spraying practices	Constant flow air line BA with full face mask	OEL x 100
Cleaning activities	automated cleaning, ↓	None	-
	vacuuming, ↓	Disposable with P2 filters	OEL x 10
	following guidelines in paragraph 73	Power assisted respirator TM2 with full face mask	OEL x 20
	↓	Powered respirator TH2 with hood	OEL x 20
		Constant flow air line BA with visor	OEL x 40
	compressed air used /operator inside the booth	Constant flow air line BA with full face mask	OEL x 100

Gloves

92 Protective gloves, boots and aprons are made from a range of synthetic materials with varying permeability to chemicals. It is likely that gloves will be needed in most situations in which coating powders are handled. The type of glove chosen needs to protect against the specific materials being handled. In most situations a good standard of impervious industrial glove will prevent skin exposure. An important consideration is the need for gloves to be electrically conducting to avoid the build-up of static when spraying. This is complicated by the fact that once the glove becomes covered in powder, it will no longer conduct the charge, the coating acting as an insulating layer. Suppliers should be consulted for advice, but this may simply require the regular changing or, where possible, cleaning of the glove. The practice of not wearing gloves or cutting holes in the palms should be avoided.

Check list – PPE

	Y	N
Have all other methods of controlling exposure been considered and implemented where it is reasonably practicable to do so?		
Is PPE required to further reduce exposure? (NB If you decide PPE is needed, you should still seek to reduce exposure by other methods to comply with the COSHH Regulations[2]).		
Has the correct RPE been selected - for dust? - for the level of exposure? - for the operators? - for the job?		
Have the correct gloves been chosen - protect against coating powders? - sufficient dexterity? - electrically conducting?		
Have workers been trained in its use?		
Are they using the PPE correctly?		
Are cleaning, storage and maintenance arrangements adequate?		

If in doubt about the correct PPE to select, contact your PPE supplier for advice.

Training

93 Workers should receive sufficient training for them to be aware of the health effects and to recognise the symptoms of exposure and the measures that have been taken to reduce the risk to their health. This health and safety training is of vital importance.

94 Training the operators on your powder coating plant is fundamental to the control of occupational exposure and to managing material costs. Without this training you are unlikely to achieve reductions in exposure and process running costs. Workers find the easiest way of doing the job that they believe will improve the finish and speed of the work. Therefore powder coating operators have been found to increase the powder delivery rate; increase the voltage to the guns; lean into the booth to spray; spray too much powder; and use brushes and compressed air to clean.

95 With increased awareness and training workers can actively help to reduce their own occupational exposure, reduce process running costs and improve quality. If you follow all the guidelines in this booklet you can reduce occupational exposure, but without commitment to training all this good work can be nullified. With this small investment in training the workers can reduce their exposure and the amount of powder used through:

- using the correct gun settings in accordance with job specifications;
- adopting the correct spraying technique;
- adopting good cleaning practices; and
- reporting defects when they arise.

96 Increased worker awareness also brings increased worker commitment as they appreciate their importance to the process, and not simply that they just spray as much powder as possible.

The more powder your operators over-spray the more money your company is wasting.

Train them and protect their health as well as saving your business money.

Case study 4

Advanced Colour Coatings (ACC) believe training is fundamental to the efficient operation of their powder coating plant. ACC apply coating powders to parts for the automotive industry and the quality of the finish is critical to their business. The company recognise the importance of training, supervision and communication for the success of the business.

Training: The company provides training for all employees to raise understanding of all aspects of the business, which includes raising health and safety awareness. Operators on the powder coating plant receive training on the technology from an external consultant. The result is that operators understand the process, follow plant set-up guidelines and strict total productive maintenance / housekeeping procedures in order to optimise powder usage, maintain quality, and follow good spraying practices. Operators do not lean into the booths to spray, consequently they get very little powder on their hands, face and overalls.

Supervision: Each job has a specification for the gun settings for the operator to follow. A patrol inspector ensures that these specifications are followed, and where they are not then corrective action is taken.

Communication: The company holds shop floor briefing meetings at the start of each day shift when operators are encouraged to make suggestions or report any problems. These, and other issues, are then discussed later in the morning by managers and supervisors with the responses and resulting actions posted on the notice board in the employees' rest room.

This high standard of training raises commitment and quality, allowing ACC to maintain contracts where high standards are demanded. Involving operators and ensuring they follow good practice reduces occupational exposure and plant running costs. Occupational exposures typically are 2 to 4 mgm^{-3} over an 8-hour shift.

Provided by Advanced Colour Coatings, Garretts Green, Birmingham.

Case study 4

Case study 4

97 This guidance contains a training module on CD, which includes narration (see inside front cover).

98 Those carrying out the exposure assessment should be trained to be able to:

- identify all the materials used at each stage of the process and how they affect health;
- assess how the workers are exposed and whether this is significant; and
- identify the measures needed to reduce exposure.

Workers at a coating powder plant used to lift their visors to see during spraying. They said that they did this as their visors were constantly being coated by powder. The supervisor explained that this happened because the workers turned up the powder delivery rate to achieve a good finish. They were worried that they would not get their bonus if they did a bad job and therefore turned up the powder delivery rate to compensate. Turning up the powder delivery rate was unnecessary to achieve a good finish.

Monitoring

99 For the purposes of the COSHH Regulations,[2] monitoring means the use of a valid and suitable occupational hygiene technique to estimate the workers' exposures to hazardous substances. These Regulations require monitoring where:

- failure or deterioration of a control measure could result in a serious health effect, either because of its toxicity or the extent of exposure, or both;
- measurement is necessary so as to be sure that an OEL is not being exceeded; and
- where necessary as an additional check on the effectiveness of any control measure.

100 If from your assessment of the work you decide that it is necessary to monitor your workers' exposure then you need to carry this out at intervals not greater than 12 months. It is likely that you will need to carry out this monitoring unless you have demonstrated that exposures are significantly below the relevant OEL. Workers on a well-operated coating powder plant may still have exposures approaching an OEL. You may only need to measure exposure to the total inhalable dust that your workers are exposed to, without the need for expensive analysis. These results can be compared annually or more frequently, as appropriate, to see improvements being made or whether control measures or working practices are deteriorating.

101 Thorough examination and testing of your LEV system as required by the COSHH regulations (see paragraph 48) will also help you to monitor your workers' exposure to coating powders.

102 There are also other ways that you can use to monitor your workers' exposure. Changes in other plant conditions may also be indicators that workers' exposure is increasing. These factors include:

- coating powder usage;
- reject rates of coated articles;
- operator appearance (ie workers getting more powder on their hands, face and overalls);
- increases in worker complaints about the working conditions; and
- observation of working practices.

103 Using these operational factors to monitor your coating plant will allow you to reinforce the training workers have received. This may reduce the frequency and extent of your air monitoring programme.

Health surveillance

104 Health surveillance is required under the COSHH Regulations[2] and is needed for:

- the protection of the health of individual employees by the early detection of possible adverse changes;
- to assist in the evaluation of measures taken to control exposure; and
- the collection, maintenance and use of data for the detection and evaluation of health hazards.

105 For workers exposed to coating powders the need for health surveillance will depend on the components in the powders in use and the degree of exposure. For agents such as TGIC it is more likely that health surveillance will be needed.

106 Biological monitoring (eg urine tests), enquiries about symptoms by a suitably qualified person (eg an occupational health nurse) and inspections by a responsible person (eg changes in skin condition by a supervisor or manager), may form part of a health surveillance programme. Biological guidance values and a detailed guide about biological monitoring are published by HSE.[4] If biological monitoring of workers is helpful then you will need a nurse or doctor to help you with the monitoring and to interpret the results. HSE's Employment Medical Advisory Service can give advice on health surveillance, including using the services of an Appointed Doctor.

References and further reading

References

1 *Chemicals Hazard Information & Packaging for Supply Regulations 1994* SI 1994/3247 HMSO 1994 ISBN 0 11 043877 9

2 *Control of Substances Hazardous to Health Regulations 1999* SI 1999/437 Stationery Office ISBN 0 11 082087 8

3 *EH40/2000 Occupational Exposure limits 2000* HSE Books ISBN 0 7176 1730 0

4 *Biological monitoring in the workplace: A guide to its practical application to chemical exposure* HSG167 HSE Books 1997 ISBN 0 7176 1279 1

5 *Seven Steps to Successful Substitution of Hazardous Substances* HSG 110 HSE Books 1994 ISBN 0 7176 0695 3

6 *Maintenance, examination and testing of local exhaust ventilation* HSG 54 HSE Books 1999 ISBN 0 7176 1485 9

7 *The selection, use and maintenance of respiratory protective equipment: a practical guide* HSG 53 HSE Books 1998 ISBN 0 7176 1537 5

8 *Triglycidyl isocyanurate (and coating powders containing triglycidyl isocyanurate) in air* MDHS 85 HSE 1997 ISBN 0 7176 1381 X

Further Reading

Working safely with coating powders INDG319 HSE Books 2000 Single copies free; available in priced packs of 10. ISBN 0 7176 1776 9

Electrostatic Powder Application - Code of Safe Practice British Coatings Federation 1997

General COSHH ACOP (Control of substances hazardous to health) and Carcinogens ACOP (Control of carcinogenic substances) and Biological ACOP (Control of biological agents). Control of Substances Hazardous to Health Regulations 1999. Approved Code of Practice L5 HSE Books 1999 ISBN 0 7176 1670 3

A Step by Step Guide to COSHH Assessment HSG 97 HSE Books 1997 ISBN 0 7176 1446 8

Monitoring Strategies for Toxic Substances HSG 173 HSE Books 1997 ISBN 0 7176 1411 5

Health Surveillance under COSHH: Guidance for Employers HSE Books 1990 ISBN 0 7176 0491 8

COSHH Essentials: Easy steps to control chemicals – a scheme to help small firms control health risks from chemicals Health and Safety Commission 1999 MISC/35/99

Code of Safe Practice: Application of Thermosetting Coating Powders by Electrostatic Spraying The British Coatings Federation 1996

Control of exposure to triglycidyl isocyanurate (TGIC) in powder coatings EIS15 HSE Books Revised 1998

While every effort has been made to ensure the accuracy of the references listed in this publication, their future availability cannot be guaranteed.

Appendix 1
Sources of advice and expertise

This guidance will help you to assess and control occupational exposure to coating powders but you may need to seek advice and expertise from other sources. These include:

- Trade Associations, who can provide advice on current practice, technological developments etc. For example:
  British Coatings Federation (Tel: 01372 360660);
  Paint and Powder Finishing Association (PPFA) (Tel: 0121 237 1121).

- Your local HSE office or HSE's Infoline (Tel: 08701 545500).

- Your coating powder or equipment supplier.

- Your PPE supplier.

- Occupational hygienists / safety consultants can provide advice on the assessment and control of exposure to coating powders. If you decide to employ the services of a consultancy then you should ensure that they are competent to carry out the work. One way to do this is to use one that is listed in the British Institute of Occupational Hygiene (BIOH) directory of Consultancies which is available from BIOH, Suite 2, Georgian House, Great Northern Road, Derby DE1 1LT (Tel: 01332 298087).

Further guidance that complements this booklet

- HSC's new guidance *COSHH Essentials: Easy steps to control chemicals* (available from HSE books) provides a structural framework which will help you do a checklist-based risk assessment and give you practical advice on how to reduce occupational exposure. Free flyer 871 from HSE books provides further information.

- The Environmental Technology Best Practice programme provides guidance on technology / techniques for profitable environmental improvement (details can be obtained on 0800 585794).

- The Department of Environment Transport and the Regions has produced a good practice guide (no 260) - *Optimisation of industrial paint and powder coating.* Available from ETSU, Harwell, Didcot, Oxfordshire OX11 0RA.

- The British Coatings Federation has produced *Code of Safe Practice: Application of Thermosetting Coating Powders by Electrostatic Spraying.*

Many other documents that may help are provided by the above government departments and by industry trade bodies.

Appendix 2
TGIC Banding limits (reproduced from Engineering Sheet No15)

HSE has now developed a validated sampling and analytical method for the measurement of TGIC (and coating powders containing TGIC) in air. It has been published as a formal reference document - Methods for the Determination of Hazardous Substances (MDHS)85.[8] The method allows employers to measure exposure to TGIC and make a direct comparison with the MEL.

An estimate of exposure to TGIC in coating powders can also be carried out by measuring exposure to total inhalable particulate (ie the total weight of material collected on the filter) and by calculating how much of this material is TGIC. In order to do this the material suppliers data sheet should be consulted to determine how much TGIC is in the coating powder. Material safety data sheets will list the amount of TGIC as being in one of two bands. These are:

Band 1: less than, or equal to 5% TGIC; and

Band 2: greater than 5%, but less than 10% TGIC.

From these it can be seen that if exposure to total inhalable particulate is below 2 mg/m^3 (8-hour time weighted average) for Band 1 powders, the exposure will also be below the MEL of 0.1 mg/m^3 for TGIC ie 5% of 2 is 0.1. Similarly, it can be seen that if exposure is below 1 mg/m^3 (8-hour time weighted average) for Band 2 powders then exposure to TGIC will again be below the MEL.

This assumes in both cases that the actual TGIC content is at the top of these ranges whereas for many powders it will be lower. In addition, HSE has established that some TGIC crosslinks with the polymer during production and therefore is not biologically available.

Taking these factors into account it is clear that, in most cases, control of exposure to total inhalable particulate from coating powders containing TGIC to the appropriate 1 or 2 mg/m^3 value will result in exposures to TGIC which are well below the MEL.

In fact, if exposure to total inhalable particulate is only just above the given values, then, in some cases, exposure may still be below the MEL for TGIC. It is therefore essential to recognise that this alternative measurement method while providing a relatively inexpensive and easier way of assessing exposure will not give the same accuracy as specifically measuring for TGIC using MDHS 85. It is an estimate or rough guide only.

In particular, this method should not simply be seen as a cheaper means of demonstrating full compliance, but more a way of gauging control or routinely monitoring exposure to assess the effectiveness of control. For example, during initial assessment both measurement of exposure to TGIC using MDHS 85 and of total inhalable particulate using simple gravimetric means could be done. Provided the results show good standards of control below the MEL, for subsequent surveys only gravimetric sampling of total inhalable particulate need be done to show adequate control is being maintained. If the total inhalable particulate levels fall further as a result of improvements then the TGIC exposure will also have fallen.

Situations will arise where workers are using a number of different powders, some containing TGIC and others not. In these cases the assessment needs to take account of the complex exposure pattern which may occur.

EH40 Occupational Exposure Limits[3] Part 3 Technical Supplement, *Calculation of exposure with regard to the specified reference periods* provides a method by which such exposure patterns can be assessed (ie to take account of periods of no exposure to TGIC).

For example, during the initial assessment where both 'band 1' and 'band 2' powders are used the assessment becomes more difficult. In these cases a sensible judgement should be made as to the significance of the total inhalable particulate measurement obtained. This involves due consideration of the relative exposure times to the 'band 1' and 'band 2' powders when interpreting the result against the 1 and 2 mg/m^3 values. Remember this method is only designed to provide an approximate assessment of exposure to TGIC so as to gauge the need for improvements. A really accurate assessment of exposure to TGIC can only be achieved by following the method described in MDHS 85.[8]